Science Fictioned

Volume 2

Lee Falin, PhD

LIGHT & LORE LLC

Also by Lee Falin

The Science Fictioned Series

Science Fictioned - Volume 1

Science Fictioned - Volume 2

Standalone Novels

Half Worlder

Science Fictioned
Volume 2

About Science Fictioned

S cience Fictioned takes ideas from cutting-edge, scientific research papers and turns them into science fiction and fantasy stories. This is the second volume in the series, and as I mentioned in Volume 1, it has its roots in two observations I made while working as a researcher in Europe:

1. There are some truly amazing scientific discoveries being made in the world these days.
2. Research papers tend to be so boring, that most people will never hear about those discoveries unless they happen to be picked up by a news outlet.

As I thought about this, I began to wonder—what if every research paper was accompanied by a thrilling story?

I published Science Fiction Volume 1—available everywhere fine books are sold—with this in mind. But alas, so far no major research journals—or to be completely honest, *no* research journals whatsoever—have yet to see the brilliance of this idea, and so those research articles continue to be a bit dry.

Undeterred, I pressed forward, sitting in my office, pouring through research articles, looking for the seeds of interesting stories. Or as a famous Pulitzer Prize winning journalist once said, "Do whatever it takes to get the story."

Editor's Note: This appears to be a quote from Lois Lane, a fictitious reporter from the Superman comics who won a fictitious Pulitzer Prize. The editorial team will do its best to warn the reader when the author is becoming a little too colorful with the truth.

I'll admit, there is a somewhat noticeable span of time between the publication of Volumes 1 and 2. But really, what's a couple of years in the pursuit of either art or science?

Editor's Note: Nearly ten years have passed since the publication of Science Fictioned Volume 1. The editorial team would like to know exactly what the author has been doing with his time since then, not to mention his royalty checks.

Regardless, today Science Fictioned - Volume 2 is here, available to readers across the world. Like its predecessor, it contains a hopefully entertaining collection of science fiction and fantasy short stories based on ideas from cutting-edge, peer-reviewed, scientific research articles.

If you're reading this and you're the editor of a major research journal, feel free to reach out. I have lots of great ideas on how to make your publication a little more interesting.

Part One
Cryptic Evolution and Ancient Warriors

The Science

Most people don't give much thought to how evolution works. In fact, aside from biologists and Pokémon fans, most people probably *never* think about how evolution works.

But just so we have some common ground for this chapter, let's imagine that one day you're in biology class, and your teacher is talking about genetics and evolution. You're sitting there, trying to stay awake, while thinking about an upcoming Pokémon battle. While part of your mind thinks about how best to evolve your favorite Eevee; another, somewhat smaller part of your brain wonders why real-life biology isn't nearly as exciting as Pokémon.

Unlike with Pokémon, in real life you usually can't see evolution happening. It's an extremely slow-moving, random force that operates on a multi-generational timescale. Evolution creates changes that can often only be detected when comparing organisms across dozens or even hundreds of generations.

The other thing about real evolution is that it's driven by a series of changes we can't see. Unlike what you sometimes see in movies, evolution starts with completely random changes to an organism's genotype (gee-no-type). A genotype is the genetic makeup of an organism. If you were to take all the DNA in one of your cells and write down its chemical sequence, that would be your genotype.

Sometimes a change to an organism's DNA results in an observable physical or behavioral change. When this happens, we say that a organism has a new phenotype (fee-no-type). A phenotype is the set of observable characteristics caused by your genotype—in other words, your phenotype is the collection of things people can notice about you because of your genes.

Things like your height, eye color, skin color, whether you are left or right-handed, whether you are colorblind, and whether you can taste phenylthiocarbamide, are all properties of your phenotype. Though I'm not sure why anyone would *want* to be able to taste phenylthiocarbamide.

Sometimes a change to your phenotype makes an organism more likely to survive long enough to pass on its traits to its children. When that happens, we call that *evolution*.

Most of the time, you can make small changes to your DNA sequence, your genotype, without affecting your phenotype at all.

Let's imagine that one bright summer morning, a mad scientist from the future teleports into your kitchen and shoots you with a random-DNA-change-o-matic laser. What would happen?

Assuming you didn't die of shock that someone from the future had teleported into your kitchen, the most likely result would be—nothing. Perhaps this would be somewhat disappointing to a time-traveling mad scientist, but the fact is that most of the time changes to your DNA—your genotype—don't do anything interesting, and your body is actually pretty good at fixing those changes.

However, sometimes small changes to your DNA can have

big effects on your phenotype—the physical characteristics your DNA controls. Sometimes those changes are good, sometimes they aren't so good.

For example, if your mad scientist friend changes enough genes related to your height, suddenly you might go from being the kid who has to stand on your tippy toes in family photos, to the one who's always picked first for neighborhood basketball games.

On the other hand, if that random-DNA-change-o-matic laser was aimed at the genes responsible for hemoglobin production, a type of protein in your blood, the result might be that you suddenly develop sickle-cell anemia—though you would also develop resistance to malaria, which might be a small consolation.

Scientists have recently (recently in evolutionary terms) begun to study another type of genetic change, which they've cleverly named "cryptic genetic variation".

While the name sounds like a bad Halloween movie, the word *cryptic* here just means *hidden*. What's interesting about this type of change is that it stays hidden most of the time, sitting in your genome, doing nothing, but then one day it can suddenly manifest in response to some environmental trigger, creating a very noticeable phenotypic change.

It's as if that slight change opened a floodgate of pent-up evolution. All of those stored-up changes to the genotype over the generations were suddenly unleashed in a massive burst of biological power.

Some scientists believe that cryptic genetic variation may play a part in the sudden rise of certain modern "lifestyle" diseases.

For example, the genetic conditions necessary for type-two diabetes may have been building up in certain family lines for centuries and are suddenly unleashed in response to the environmental effects of poor diet and lack of exercise.

Or perhaps the sudden rise in coeliac disease has been building up in our genomes for generations, only to be released now in response to recent changes in agricultural practices and pesticide use.

Or maybe both of those diseases are simply a result of spending most of our day sitting, typing on a keyboard, and eating fast food. Perhaps only time-traveling mad scientists from the future will ever know for sure.

Whether or not all lifestyle diseases are a result of cryptic genetic variation, if you're a writer, when you hear the term *cryptic genetic variation,* your mind starts to drift towards gothic fantasy.

What would happen, the writer's mind asks itself, if someone from modern times possessed the genetic makeup required to channel the ancient combat magic used by the warrior-kings of old?

What if such powers had been passed down throughout the generations, hidden within our genomes, until one day, those latent abilities were unexpectedly triggered by a sudden genetic call-to-arms?

The Fiction

The Journal of Dr. Thomas Burbank
3rd of August, 1848

I arrived early this morning at the British Museum. The university provided a most excellent room for me not three blocks away from the main entrance. Light fog and the accompanying London chill hung in the early morning air, though it could do little to sink my buoyant spirits. I'm still rather taken by my good fortune to have received a private invitation from Sir Layard himself. I must remember to thank Lord Canning for having made the introduction.

When I arrived at the museum grounds, I found the place quite deserted. At first, I'd worried that I'd come on a day the museum was closed, but I soon came across a team of workmen who were constructing a new exhibit. They explained to me that most of the museum staff was on holiday and one of them kindly escorted me to the site manager, a Mister Henry Briggs.

I found Briggs to be most knowledgeable regarding not just the work on the new exhibit which he oversaw, but also as related to the general history of the museum and the local environs of London. As we wound our way through a maze of laborers already well into their work, he called my attention to the great quadrangle building with its awe-inspiring Greek facade, which was nearing completion.

It was as though a small piece of ancient Greece had been dropped into the center of London by one of the Olympians of old. I must remember to ask if it would be at all possible for me to take a tour of that building while the museum is still closed to the public.

I must admit to feeling a sort of boyhood thrill whenever I encounter a great museum. Their ability to thrust the stories of antiquity into our modern lives has an almost mystical appeal.

After showing me around the grounds, Briggs led me into the museum's main building through a side entrance, escorting me to a large gallery where a variety of tapestries, paintings, and other ancient artifacts were on display. Rays of dust-flecked sunlight streamed through the eastern on the east side of the gallery, which only added to its sense of majesty. It was then that I got my first look at Sir Layard in person.

He was a striking figure of a man, with a full head of thick black hair, which appeared to merge seamlessly into his bushy beard and mustache. His tanned, weather-beaten complexion indicated one who had spent long hours toiling in the hot sun and dusty winds that always exist in foreign archeological digs.

Not for the first time, I wondered whether it was good fortune or ill that my own field of expertise in Celtic antiquities usually kept me confined to the gentler climes of Wales and Ireland.

When we entered, Layard was stooped over a display table, examining what appeared to be a long strip of dark metal. He glanced up at our approach, then beamed, striding towards us with a hand extended in warm greeting.

"Ah, Dr. Burbank, it is so good to finally meet you in

person," he said. "Your recent paper on the shamanic rights of the ancient Britons was most excellent."

"Thank you," I replied, in what I hoped was an appropriately demure manner. "I assure you the honor is entirely mine. I am most eager to learn more about your work amongst the Ottomans."

Layard nodded a dismissal to Briggs and passed him a half crown. The site manager gave a short bow and left us alone in the vast hall.

"Ah yes, the Ottomans are a fascinating people to be sure," Layard said as he led me over to the display he'd been examining. "But it is my more recent work in the land of Assyria for which I need your assistance. I found something there I hope will be well within your area of expertise."

He gestured towards the table. A single-edged knife—or seax—nearly two feet in length lay there, its black iron finish seeming to absorb the morning sunlight. A deep groove ran down its center. A double line of intricate runic glyphs, wrought in copper and silver, ran along either side of the groove.

A chill ran down my spine as I gazed upon the artifact, the sort one often feels when confronted with such a well-preserved piece of antiquity. "You found this in Assyria?" I asked.

"In a tomb just outside of Mosul," Layard said. He pointed towards the runes. "Can you translate the inscription? I thought at first, they must be Old Italic in origin, but there are some runes I cannot identify. You are welcome to handle the blade in order to get a closer look."

I sat down my case and took the blade reverently in both hands. No sooner had I done so, when I cried out at a sharp pain in my right thumb.

I suppose in my excitement I had been careless and pricked my finger on the edge of the blade. To my embarrassment, a small drop of blood ran down my thumb onto the artifact, seeming to almost shimmer slightly against the blade's dark surface.

I began to offer my sincerest apologies to Sir Layard for my clumsiness, but he waved my apologies away, seeming genuinely unconcerned that I'd just stained a priceless historical artifact with my own blood. He is a most capital fellow.

"But can you make out the runes?" he asked, nodding towards the blade.

Forgetting my thumb for a moment, I scrutinized the double line of writing, taking care not to let more blood drip onto the surface.

"They appear to be Anglo Saxon. See here," I said, gesturing to the first line of runes with my good thumb. "These are nearly identical to the runes in the Vienna Codex, but some of the characters near the end have been transposed. See these last two characters? In the Vienna Codex, we have *ea* followed by *y*, which we believe represent death and water respectively. But on this blade, the characters are transposed. It is most curious."

"Indeed," Layard said with a faint smile, "Perhaps the forger of the blade wanted to keep death on the far side of the water then?"

I chuckled at his dark humor. "Quite. Though some scholars believe *y* is better translated as 'wild ox', so perhaps it is some sort of ox that is keeping death at bay."

"And what of this second line of runes?" Layard asked. "Are those part of the Vienna Codex as well? More tales of formidable oxen perhaps?"

I shook my head, "I don't believe so. It's most likely the name of the bladesmith, or perhaps of the owner of the blade. It appears to read as *Be-ag-noth*."

Just as I uttered the transliteration, a slight tremor shook the room. Flakes of ceiling plaster rained down upon us amid a light haze of dust. I eyed the ceiling uneasily.

"Pay no mind, Dr. Burbank," Layard said. "Those tremors have been happening regularly of late—a result of the construction work being done on the new exhibit. I'm half-convinced that in their effort to add a new section to the museum, our workmen will inadvertently destroy one of the existing sections."

I chuckled uneasily at this, eyeing the ceiling. But his calm manner reassured me, and I asked if I might make a quick sketch of the artifact and its inscriptions for later study. He assured me that nothing would please him more.

He then apologized that he had an urgent appointment that afternoon and asked if I might dine with him the following evening at his private apartment in Westminster. I agreed most readily as I had many questions I longed to ask him about his work abroad.

∾

The Times Morning Edition
5th of August, 1848

Residents of London living proximate to the British Museum have once again voiced their concern regarding—in the words of one local resident—"The blooming noise and other devilry that's been goin' on all hours of the night."

Though the site manager in charge of the works, a Mister Henry Briggs, stated quite emphatically that work on the grounds is stopped each day well before the dinner hour, several local residents reported that late into the night on the 4th of August, a series of crashes, the sound of breaking glass, and deep rumbling sounds could be heard as far away as Westminter. One resident described the commotion as being "as if the devil 'imself was loose in the museum."

~

The Journal of Dr. Thomas Burbank
5th of August, 1848

This entry will be somewhat shorter than usual, as I find myself terribly done in this evening. The beastly cut on my thumb aches like anything, making writing most difficult. Though I wasn't conscious of it at the time, the pain must have interfered with my sleep as well, because I've been on the point of exhaustion most of the day.

I dined with Sir Layard as planned, though I fear I may have offended him, as I began to nod off several times during dinner. This was most embarrassing, especially as I am genuinely curious about his discoveries abroad. I wonder if he noticed however, as he kept bringing the conversation around to my own history, encouraging me to talk—not just about my research and academic career, but also inquiring into my home life and even my family's ancestry.

"I believe you mentioned in one of your lectures," he said after we had retired to the library, "that you are a direct descendant of Eadred of Wessex, the former King of Northumbria, did you not?"

I nodded, wondering not just at his attention to detail, but also curious as to when he'd ever been in a position to hear one of my lectures. "That's the way my grandmother told the story anyway," I shrugged. "According to her, our lineage is the noblest possible mixture of all seven royal lines of the Heptarchy."

"That is most satisfactory," he said, lips twitching slightly.

I nodded uncertainly, unsure what he found so amusing, then recollecting myself I excused my stupidity, blaming my confusion on how devilishly tired I felt. Rather than taking offense at my lack of manners, Sir Layard showed his in excess by inviting me to lodge with him for the evening.

"I have plenty of rooms here," he said. "And most are almost always unoccupied."

I accepted most gratefully, as the idea of traveling back to my own lodging seemed almost insupportable in my present state of fatigue. I'm now resting comfortably in a spacious room on an upper floor. Though I'm afraid I haven't the will to describe its furnishings in detail, as I am so terribly tired.

~

Extract from Sir Layard's Personal History
5th of August, 1848

I know this volume of notes can never be shown to the public, but I keep my record here in the hopes that perhaps some future descendant, or angel of heaven will come upon it and know the steps I have taken to safeguard the kingdom from the fiends of hell. Hopefully they will judge me by my motives rather than my actions, knowing that at least the former were pure and unselfish.

Had there been any other path open to me, I would have pursued it. But my ancestry combined with my own experiments makes it clear that I am an unsuitable host for what is needed.

If there had been a way to secure Dr. Burbank's permission for what is required of him, I would have done so. But even had our friendship been at such a depth that I could ask such a think

—and even had he been a schoolboy chum reared together with me in the same household, I would still hesitate—how unlikely is it that he would believe half of what I would have to tell him in order for him to make an informed choice on the matter.

But these thoughts are pointless. The completed rites have already set the process in motion, and no power from hell or earth can stop it now. May the God of Heaven watch over the good Dr. Burbank, and may both of them have mercy upon my soul.

∽

The Times Morning Edition
6th of August, 1848

Late last night, multiple witnesses reported a strange disturbance along the banks of the river Thames in Westminster. The facts of the matter are still unclear, partially due to the fact that most of the witnesses had been heavily drinking at the time.

However, a constable who asked not to be named, claimed to have seen no fewer than three-score "creatures of a most unearthly description" prowling the banks of the river.

Other witnesses reported seeing all manner of unearthly fiends and creatures rising from the muddy banks, from half-rotten corpses to reanimated skeletons. The so-called "monstrous horde" was reported to make its way towards Westminster Abbey, but was harried by what one witness described as "a bloomin' giant—in armor and all, wielding a massive iron sword like a cricket bat."

Upon visiting the scene of the alleged battle in the early hours this morning, this reporter could find no trace of a cricket playing giant, nor any sign of spirit or monster, other than the usual complement of mudlarks and vagabonds one normally encounters around the Thames in the early morning.

Whether this report is indicative of a resurgence in the activities of a group of the so-called resurrection men, or merely revelatory of the poor quality of Westminster brandy, we may never know.

∽

The Journal of Dr. Thomas Burbank
6th of August, 1848

I hardly know how to begin my recollection of last night, or even if I should put down my thoughts in writing. Surely if someone were to read this account, they would think me mad.

After retiring to my room in Sir Layard's lodgings, I fell asleep quickly, but my dreams were troubled. I found myself rising from bed just after midnight. It was the oddest feeling, for I found myself striding purposefully from the room, along the passage to the stairwell, and then down the stairs and out of the house without any conscious thought or will behind my actions. It was as though I were a prisoner within my own body, forced to witness as some other force controlled my movements.

Though that experience on its own should have been enough evidence to convince me I was dreaming, the next sequence of events made it a certainty. As I strode across the lawn, I saw Sir Layard kneeling in the moonlight, holding the iron blade from the museum in his outstretched hands, as if in offering to me.

"Take this blade mighty Beagnoth," he said, "and once more defend this realm from the unholy forces of evil that threaten it."

As I reached out and gripped the hilt of the weapon, the runes along its edge flared brightly and I felt a surge of power and purpose wash over me.

Layard was now shrinking away from me as if I were floating upwards from the ground. It was then that I realized Layard wasn't shrinking, nor was I floating. I was rapidly swelling in stature, growing larger by the second, and I somehow knew instinctively that this change was connected to the surge of power coming from the blade I now gripped in my hand.

As the rate of growth ebbed, plates of black iron began shimmering into existence around my body, forming a great suit of mail. By the time the entire process had completed, I found I had grown to a height of nearly two stories, and the knife in my hand had grown proportionally and was now the size of a massive broadsword.

I then felt an inexorable pull urging me in a southerly direction, towards the river Thames. I began to run, nearly as fast as some great steed, and within minutes I stood in the shadow of the spires of Westminster Abbey.

My attention was drawn to a movement near the banks of the Thames. What appeared to be a group of men surged about, seeming to claw their way out of the muddy banks alongside the river. Whatever force controlled me urged me towards them, and I proceeded at a run.

As I moved closer, the light of the moon showed me that they weren't men at all, but horrible skeletal warriors and monsters from the legends of old. They wielded a variety of rusted and broken weapons. A manic red light, like the very fires of death, shone in their eyes.

Perhaps strangest of all, I felt absolutely no fear, but was instead overcome by an intense hatred towards these creatures.

With a bellow of rage and power, I swung my sword, cutting through their ranks like a farmer wielding his scythe at harvest. Soon, the banks were clear of the fiends, and I found myself longing for more foes to vanquish. It was then that I heard Layard's voice behind me.

"Mighty Beagnoth," he shouted. "You are once again victorious against the hosts of evil. Your enemies are no more. Please lay down your blade and release your host until honor once again calls you forth to defend the realm."

My grip on the blade tightened. I found myself enraged that so piteous a man would dare command me—the mighty Beagnoth, champion of the Heptarchy. I raised my weapon, prepared to strike him down.

My conscious mind felt an immediate horror at this thought, and I found myself once more in partial control of my body.

I stood as if frozen while poised to strike, as my own will battled with that of Beagnoth. In a final, desperate surge of will, I forced my arm to flick forward and release the blade. It hurtled end-over-end through the night until it landed with a soft splash in the waters of the Thames.

The power of Beagnoth left my body as suddenly as it had come upon me, and I began to rapidly shrink back to my normal stature. I fell forward, collapsing onto my knees at the feet of Sir Layron, and was struck by a sudden coldness that seemed to sap the rest of my strength.

Layron helped me to my feet and wrapped me in some kind of dressing gown. "You have done well Dr. Burbank. Truly, the blood of the warrior kings of old runs pure in your veins. Your ancestors would be proud."

That's where the dream ends, though once again I find myself horribly tired this morning.

~

The Times Morning Edition
22nd of May, 1857

" *The curator of the British Museum announced today that an ancient artifact of Anglo Saxon origin has been discovered by the site foreman, a Mister Henry Briggs, on his morning walk.*

Mr. Briggs was walking along the Thames on his way to visit his mother in Battersea when he noticed a large iron blade protruding from the river's muddy banks.

Thanks to the runic symbols engraved along the blade's edge, museum staff have identified the artifact as a seax, an ancient single-edged knife used by Anglo Saxon warriors.

The exact origins of the blade are still being researched, and though this reporter reached out to two different experts in Anglo Saxon antiquities, Dr. Thomas Burbank and Sir Austen Layard, both men declined to comment on the discovery.

Part Two
Solar Scientists and Vintage Glory

The Science

One of the most fascinating things about outer space is that there are so many things just floating around out there, hidden in plain sight.

In 1610, Galileo famously used a telescope to make the very first observation of Jupiter's four largest moons—Io, Europa, Ganymede, and Callisto. This was a groundbreaking discovery at the time, and helped reshape our understanding of the solar system.

In modern times, anyone with a halfway decent pair of binoculars can see the Galilean moons for themselves. And that's what's so fascinating. The moons have always been there —in human terms, just waiting for someone to notice them.

Even our own sun, which we see quite literally on a daily basis, still has secrets and mysteries just waiting to be noticed, especially below its surface. Unfortunately, it's painfully difficult to observe the sun, especially the stuff below its surface. So one day, a group of scientists got together to try and figure out what to do about this.

Their first order of business was to decide what to call themselves. Scientists who study dinosaurs are called paleontologists. Those who study the earth are called geologists. But what should scientists studying the sun be called? Sunologists? Solargists? Suggestions were made, votes were cast, focus groups were polled, but there was no clear winner.

Then one day, one of the scientists was bragging about a trip to Greece he'd recently gone on for a conference—scientists always manage to hold their conferences in exotic locations near picturesque beaches—when one of his fellows leapt to his feet and proclaimed, "That's it! We'll call ourselves helioseismologists, after Helios—the Greek god of the sun."

The other scientists thought this was a great idea, and from that time forth, scientists that studied the structure and dynamics of the sun were known as helioseismologists.

Editor's Note: While these scientists are in fact called helioseismologists, the authenticity of this naming story is highly doubtful. Or as one of our teenage interns put it— that's cap and totally sus.

With that important matter worked out, the scientists set to work figuring out how best to study the mysteries of the sun's behavior—specifically how it behaved beneath its surface. As they felt that these under-surface dynamics would hold the key to many interesting discoveries.

In 1995, helioseismologists at the European Space Agency and NASA teamed up to design and launch the Solar and Heliospheric Observatory (affectionately named, SOHO), a satellite designed to hang out around the sun and provide helioseismologists the type of data they love to study.

Aside from the fact that it was going to be hurtling through space, SOHO was also going to be in relatively close proximity to the sun—roughly 1.5 million kilometers closer to it than the Earth is—so it needed to be built to last.

Even though its mission was originally planned to last just two years, at the time I write this, it's been plugging away in space for nearly thirty years. Meanwhile, my four-year-old smartphone has a bit of a meltdown every time I try to take a photo. Smartphone factories really need to start consulting with space engineers about their design practices.

Over the years, SOHO has made some fascinating discoveries, not just about the sun, but about a host of other things that happen to be floating around in space, especially comets. As of 2024, apart from a ton of solar data, observers have used SOHO's instruments to discover over five-thousand comets.

However, one of SOHO's most important functions is to help scientists predict when a solar storm, or coronal mass ejection (CME), might occur, as they can produce massive solar flares. While scientists still don't understand exactly *why* CMEs happen, since solar flares can cause problems with Earth's power and communication systems, it's important that we know *when* they will occur.

The largest CME on record was in 1859 and was later named "The Carrington Event", after the British astronomer who first recorded it. The day Carrington recorded the CME, another astronomer from Scotland, Balfour Stewart, reported the effects of a massive geomagnetic storm.

While all signs pointed to the fact that these two events were likely related, Carrington simply shrugged and said, "One swallow doesn't make a summer." He then sat down for tea.

Between Carrington's casual use of British idioms and Stewart's thick Scottish accent, astronomers in other countries were having a difficult time understanding this strange phenomenon.

Meanwhile, American telegraph operators had more important things to deal with than the nuances of cross-cultural language barriers, namely that many of their telegraph systems had started shorting out, emitting sparks, and bursting into flame.

If that weren't troubling enough, some telegraph operators noticed that after hurriedly disconnecting the batteries from their systems, they were still able to send and receive messages for hours. Some of them noted that during this geomagnetic storm their systems seemed to be working more efficiently than ever.

It was as if after turning off your smart phone, you discovered that it could still send texts, and somehow all your green bubble friends had become blue bubble friends.

In more recent times, solar storms have caused damage to satellites, shut down power grids, and resulted in disruptions to various digital networks, including those that control stock market trading.

A solar storm in the early spring of 1989 disrupted the power grid in Québec, causing a massive power outage. While power was eventually restored, spending nine hours dealing with the extreme cold of Canadian winter left more than a few residents wishing for summer, no matter how many swallows were involved.

Frightened at the thought of having to spend another winter night in Canada without heat, and even more frightened at the thought of the stock market being disrupted, politicians and business leaders worked together to make some changes to our electrical networks.

In the last few decades, power grids around the world have been upgraded to help them better handle geomagnetic storms caused by solar flares. This was made somewhat easier with the help of SOHO, whose early warning systems can sometimes give earth a couple of hours' notice that a solar storm is imminent.

In our modern, digitally saturated world, even a couple of hours warning can be useful, as a truly massive solar storm—the big one, as some helioseismologists refer to it—could completely disrupt global power and communication networks, leaving people without a way to communicate—or at least *most* people.

The Fiction

Dr. Henry Lewis—Hank to his friends, tore out another page of the research journal he'd been reading, crumbled the page up into a tight ball, and tossed it towards the trashcan in the corner. The crumpled paper bounced off the edge of the can and fell to the floor, joining several other wads of paper.

"Played basketball in high school, huh?" Anne asked.

Hank flushed. He hadn't realized she'd been watching him. Lately, it seemed like Anne was always watching him—it made him nervous. He supposed it could have all been in his head—it wasn't like there was much else to look at in their cramped, windowless office.

Their desks were arranged back-to-back; the only way two desks could fit into the tiny room. Hank always thought it ironic that Palomar Observatory's only helioseismologists had been crammed into an office with no access to sunlight. His supervisor had said it had something to do with budget cuts and internal politics, and had assured them it was just temporary. That had been five years ago.

"Just a little precision verses accuracy experiment," Hank said offhandedly. Anne was a few years younger than Hank, but these days it felt like everyone at the observatory was younger than Hank.

Anne leaned around her monitor to consider the wads of paper scattered haphazardly around the base of the trashcan. "I'm having trouble finding either precision or accuracy here." She bent down and picked up one of the wads of paper, unfolded it and scanned the text. "I see you're putting the pages from Dr. Marsh's latest research article to good use."

Hank scowled, tearing out another page. "I thought it might make me feel better."

"Is it helping?"

Hank crumbled up the page and tossed it across the room. It bounced off Anne's leg and fell to the floor. He sighed. "Not really, but back when I played on the Olympic team, the hoops were larger."

"Right," Anne drawled. "That was with the original team from Athens?"

"Funny." He flipped the journal over, scowling at the picture of Marsh on its cover. In the photo he was learning against the wall of the observatory, attempting to look stoically towards the heavens—the pretentious weasel.

Hank snatched the journal off his desk, brandishing it angrily. "Do you know that this is the third time, Anne! The third time in as many years that the weasel has stolen *our* data to use in his paper."

"Technically," Anne began, "we signed a collaborative agreement when we joined the observatory. So, any scientist's data is available for use by the rest."

Hank opened his mouth to protest, but Anne held up a finger. "And he *did* credit us in his last article."

Hank snorted, reciting the passage in a pompos, nasally voice "I'd like to thank my colleagues for their valuable input and helioseismic data."

Anne laughed. "That does sound just like him."

Hank ground his teeth. "He didn't even mention us by name. Just *once*, Anne, I'd like us to get credit for our work, just one time I'd like to see my name in print. It's been nearly five years—"

"I know," Anne soothed. "But at least—"

She broke off as a light started flashing on her console. She glanced down at it, then looked at Hank, her eyes wide. Hank stared at her, seeing his own shock mirrored on her face.

"Tell me that's not—" he began, then jumped as a buzzer sounded from his own computer. His hands seemed to move of their own accord as they typed out a sequence he'd used thousands of times—perhaps tens of thousands. But this time...

"Hey, hey!" Anne said, her fingers flying across her own keyboard. "Who cares about Marsh's paper when you can see the beautiful mountain-like peaks of these perturbation graphs? This could be it Hank, your big break!"

"*Our* big break," Hank replied automatically. "Do you have SOHO telemetry online?"

"Of course."

Hank scanned the readings scrolling across his screen. The graph for the seismic perturbation index *did* look like a mountainside. A *steep* mountainside. "Are you seeing this?"

"The travel-time perturbations are off the cart," Anne said breathlessly. "Think we ought to phone this one in?"

Hank looked up at her, then his eyes trailed towards the dark red emergency phone that sat atop his desk. A thin layer of dust covered its buttonless surface. They had never phoned in a solar flare warning before. As far as he knew, nobody had phoned one in for at least as long as he'd been at the observatory —maybe even longer. There'd never been one big enough.

His eyes flickered back to his screen; the perturbation index was still climbing. He swallowed, stretching his hand towards the phone he'd never used, then hesitated. If they were wrong...

Anne started at him, eyes wide as his hand hovered above the phone. Her eyes flicked back to the screen. "Readings are still increasing," she whispered.

Hank took a deep breath, set his jaw, and picked up the phone. His hand trembled slightly as he held it to his ear.

Just over twenty-four hours later, Hank sat in the observatory's large conference room, readjusting his papers for the tenth time. The room was one of the newest additions to the observatory. The western wall was covered in floor-to-ceiling windows. The setting sun shone brightly and likely would have made the room too hot to use if the windows weren't coated in polarized shielding. Hank glanced down at his charts, reflecting on the irony of giving this presentation in a room protected by solar shielding.

Anne dropped into the chair next to him. "Where were you all night?" she asked. "I tried to call you three times."

"I was out walking," Hank said. Walking at night always calmed his nerves. Maybe it was the fact that he spent all day at work studying the sun.

"One of these days you really need to invest in a cell phone," Anne said. "It would make it a lot easier for people to reach you."

"My old landline works just fine," Hank said. "Besides, one of the reasons I go out walking is so people *can't* reach me. Carrying a cell phone would defeat the purpose."

Hank looked up from his notes and scanned the room, it was nearly full now. Dr. Marsh sat near the far end of the table, regaling the men nearest him with a story about how he'd recently secured a massive grant from the Department of

Defense. Hank gritted his teeth. That grant had been secured using data he and Anne had recorded, and it gave Marsh enough funding for a state-of-the-art lab and a sizable team of research assistants. Meanwhile, he and Anne shared what was essentially a storage closet.

Marsh looked up just then, catching Hank's eye. His face broke into a knowing smile. He said something in a low voice to those nearest him. They chuckled, a couple of them casting sidelong glances at Hank.

Hank flushed and looked away. On the other side of the room, two military officers sat near the head of the table speaking in low voices with Dr. Bowers, the head of the observatory. Bowers had been a great research scientist in his time, but these days he spent most of the day buried in administrative duties. Hank shuddered at the thought. Even if he and Anne were crammed into a broom cupboard, at least they were still directly involved with research.

The only other person in attendance was a young woman Hank didn't recognize.

"Who's that woman?" he hissed.

Anne glanced up. "That's Charlotte O'Neele."

Hank stared at Anne blankly. "Who?"

Anne rolled her eyes. "From Nation's News Hour? Don't tell me you don't have a television either."

"I have a television," Hank said. "I just don't use it very much." The last thing he wanted to do after staring at telemetry reports all day was go home and stare at *another* screen. "What's she doing here?"

"Covering the story," Anne said. "Apparently, she flew in with the military officers. I'm sure she'll want to interview you after the meeting."

"Me?" Hank asked, genuinely shocked.

"Of course," Anne said, grinning. "You're the head helioseismologist, and this is huge news."

Hank flushed. As much as he wanted to be recognized for his work, he had expected that would come later, after he'd thoroughly reviewed and published his research. He never expected to be on the news.

The woman looked up, and Hank realized he'd been staring at her.

She winked at him, flashing a dazzling smile. Hank felt his cheeks grow warm and dropped his eyes back to his notes, surreptitiously wiping his sweaty hands on his pants. He was pretty sure the woman was laughing at him.

Relax, he told himself. *You just need to stand up, deliver the message, then sit back down again.* Hopefully there wouldn't be any questions. His eyes flicked to Marsh and his cronies, and his heart sank. Knowing Marsh there *would* be questions—probably unpleasant ones.

Hank jumped as he felt someone lay their hand on his arm. Anne leaned in close on the pretense of examining one of his charts. "Relax," she whispered. "You've given hundreds of presentations on solar phenomenon."

"Right," Hank hissed. "To other astronomers and astrophysicists. And you know Marsh is going to try and tear our data apart."

Anne rolled her eyes. "Marsh wouldn't know real science if it smacked him in the face. Actually," she mused, "I wouldn't mind seeing that."

"Seeing what?" Hank asked, confused.

"Marsh being smacked in the face."

Hank chuckled, and his stomach unclenched slightly. Anne had a knack for making him feel better. One of these days he really should get around to—

"I think we're all here, Hank," Bowers said from the head of the table. He glanced up at the large wall clock. "You can start whenever you're ready."

Hank looked up and regretted it almost immediately. All eyes were fixed on him. He shuffled his papers, stalling as he tried to remember the opening line he and Anne had practiced. It hadn't been funny—Hank hated it when people started presentations with jokes—but it had been witty.

Someone coughed softly.

Hank felt a stab of panic.

Anne squeezed his arm.

Hank took a deep breath and got shakily to his feet, gripping his slideshow remote control like a lifeline.

"Hello," he croaked, then cleared his throat and tried again. "Hello, I'm Dr. Henry Lewis—but you can call me Hank. I'm a helioseismologist here at the Palomar Observatory." Why was it so hot in here? Part of his mind wondered if the polarized shielding on the windows needed to be reapplied.

Anne kicked him under the table.

He winced, then clicked the button on the remote that would advance the presentation to the next slide. The presentation shut off.

Hank frowned at the remote. "Um...sorry, I—"

Marsh chuckled softly. "You'll have to excuse my colleague," Marsh said to the room at large. "He doesn't even own a cell phone, prefers a landline."

"Really?" Charlotte O'Neeele asked, seeming genuinely interested. "That's fascinating."

"It's certainly something," Marsh said, suppressing a laugh.

Hank scowled, but Anne just plucked the remote from his hand and brought the presentation back up, advancing it to the next slide.

"Thank you," he said weakly, keeping his eyes focused on the presentation screen. "A few hours ago, my colleague, Dr.

Anne Evans, and I observed an intense spike in travel-time perturbations at a depth of approximately sixty-five thousand kilometers into the convective envelope of the sun, as shown on this chart."

Hank paused, turning to face the room as he let those numbers sink in. That hadn't been *the* line, but he thought it was pretty good. Marsh watched him intently, his eyes narrowed slightly. Bowers looked shocked. Everyone else looked slightly puzzled.

"And that chart—it's to scale?" Bowers asked, alarmed.

Hank nodded. "Yes, I'm afraid so."

"And you've cross-referenced those readings?" Marsh asked pointedly. "You're sure this isn't a result of interference from the ionosphere—again?"

One of the astronomers seated near Marsh chuckled softly.

Hank glared at Marsh, resisting the urge to throw the remote control at his face. *One* little mistake resulting in a retraction of his last paper—had it really been ten years ago—and Marsh still found a way to bring it up every chance he got. Well, once he and Anne finalized this data and submitted it for publication, things would be different. They would finally get credit for their work, and then maybe—

Anne cleared her throat, pointedly looking at Hank.

"Ah, yes..." he said sheepishly. "We cross-referenced the readings with the Kitt and Lowell observatories and they concur with our analysis," Hank said, fighting to keep his voice calm. "And we have a meeting in a few hours with a team of astronomers from the Royal Observatory in Greenwich to discuss a joint emergency press briefing on the findings."

"Really?" Marsh asked, clearly skeptical. "Who's your contact there?"

"We spoke directly with Dr. Fran Elliot," Anne said casually, eyes on Marsh. "I believe you worked with her at some point, didn't you Dr. Marsh?" Anne pretended to look thought-ful. "Or...is she the one who kicked you off her research team after you—what were the exact words she used? Oh yes, stabbed her in the back."

Marsh flushed, opened his mouth to reply, then snapped it shut and sat back in his chair, arms folded, glaring at the presen-tation screen.

Bowers cleared his throat. "Perhaps we should continue with the presentation."

Hank suppressed a grin, his courage bolstered by the sight of Marsh being cowed.

He nodded to Anne, who clicked the remote, bringing up the next graph.

"As you can see," Hank said, gesturing towards the screen. "The exponential change in seismic perturbations shows us—"

"Excuse me, Doctor."

Hank turned, the older military leaned forward, a frown on his stern face. Hank didn't know his exact rank, but judging by the sheer volume of medals pinned to the front of his uniform, it must have been pretty high. "I'm sorry to interrupt," he said, "and I'm sure your charts are impressive, but it's been a few years since my uh—high school science days. Maybe you could give us a quick summary—in plain English."

Bowers grimaced apologetically. "Hank, this is General Hobbes. He's the head of the president's military science task force."

Hank nodded, absently wondering why the head of any *science* task force would have such a poor grasp of basic astro-nomical theory. He glanced back at the chart showing the perturbation graph. If the general couldn't understand the data in such a basic graph, Hank wasn't sure how much simpler he could make this.

"The p-waves," Anne whispered.

Hank nodded thankfully. "So, we measure the p-waves," Hank began, then stopped as the General raised an eyebrow. Hank sighed and tried again. "That is to say, we measure the seismic activity of the sun using an indicator called the perturbation index. A spike in the perturbation index means that solar flare activity is set to increase."

General Hobbes frowned. "A solar flare?"

Hank nodded, "Correct. Though technically the readings predict solar storms rather than flares exactly. According to satellite telemetry, a massive solar storm is eminent, and a solar flare is likely to—"

The general held up a hand, cutting him off. "Now this could be my ignorance showing again, but could one of you explain why I had to wake up at three AM and fly all the way to California, just to hear about storms on the sun?"

Hank stared at him, dumbstruck. "Well...as I said, it's quite large, and..." he turned to Anne again for help.

"What my colleague is trying to say," Marsh interrupted before Anne could speak. "Assuming his readings are correct of course—is that we've detected an eminent astronomical event that presents a significant threat to Earth."

Hank scowled, feeling his face redden. Did he just say *we?*

"What kind of a threat," the general asked, turning to face Marsh.

Marsh stood, walking to one of the whiteboards that lined the wall behind him. "The strongest solar storm on record to this point was in 1859—the Carrington Super Flare. The levels of electromagnetic radiation were so intense that telegraph wires around the country shorted out, starting several fires. In other places, telegraph operators were still able to send signals over the wires, even though the power had shorted out."

"You mean the radiation was powering the transmissions?" the younger military officer asked.

"Exactly," Marsh said. "Now fast-forward to today," Marsh continued, speaking quickly, all eyes on him as began drawing a hasty illustration of a solar flare striking the earth.

Hank dropped slowly into his seat, shaking with barely suppressed anger. How dare that insolent weasel hijack his presentation like this? For crying out loud, he wasn't even a helioseismologist!

Anne put a hand on his arm. "Easy, this is still *our* data..."

"Like that's ever mattered before," Hank hissed.

"—which makes every internet router, cell phone tower, and power station vulnerable," Marsh continued. "Once the flare hits, power and communication could be out for days—maybe longer."

Hank scanned the room, feeling his heart sink. Marsh had their complete attention. The presentation might as well have been his from the start.

"So, how does this storm compare with the one in 1859?" The general asked.

Under the table, Anne kicked Hank again.

"Ow!"

Everyone turned to look at him.

Anne looked at him meaningfully. But before he could say anything else, Marsh was using a laser pointer to indicate the relevant section of the graph.

"According to satellite readings, we estimate the coming storm to be about a thousand times the magnitude of the one in 1859—and it hits in less than twenty-four hours."

Hank opened his mouth to speak but broke off as the military officers got to their feet.

"If you'll excuse us gentlemen, if the nation is about to lose power and comms, I need to alert the President."

Marsh jumped to his feet, intercepting them at the door and extending a hand. "My card General, in case I can be of any assistance in your briefing with the President."

The general nodded curtly. "We'll be in touch," then strode from the room.

"I'll take one of those too please," the reporter said, darting over to take one of Marsh's cards. "I've got to phone this in while I still can."

"Of course, and if you need me to answer any questions..." His voice trailed away as the two of them left the room together, followed by the rest of the astronomers.

Hank stared opened mouth, feeling like he'd been hit by a whirlwind. "He did it again...that fink."

"I don't believe it," Anne said, shaking her head.

Hank's sighed, gathered up his notes, and walked slowly towards the door, shoulders slumped. In that moment he felt every one of his fifty years.

"Where are you going?" Anne asked, her temper rising. "Aren't you going to do anything about this? You can't just let Marsh get away with this again!"

"Yes, I am going to do something," Hank said quietly, his voice heavy with defeat. "I'm going home."

Two days later, Hank was sitting at his small kitchen table, eating a grapefruit. He didn't usually eat grapefruit without toast, but since the power had been knocked out by the flare, he had no way to run his toaster.

He sighed as he took another bite. His back was sore, and his head ached slightly. He hadn't slept well last night. He couldn't get the image of Marsh out of his mind. He kept going over that disastrous presentation. He should have practiced it like Anne had suggested, but there hadn't really been time.

Maybe if he'd been more confident...he shook his head. Marsh still would have found a way to hijack the meeting. Maybe it was time to retire. He could sell his house, maybe travel a bit.

He shook his head. He loved research too much to retire this early, even if he hadn't managed to publish anything in nearly ten years. If only—

Hank started as his phone rang. Was the power back on already? He jumped up and hurried over to the phone. It had been hanging on the wall of his kitchen since he'd bought the house, nearly twenty-five years ago. He picked up the receiver.

"Hello?"

"Is this Dr. Henry Lewis?" The voice was masculine and sounded familiar, though Hank couldn't place it.

"Yes, I'm Dr. Lewis. Who am I speaking with?"

"Thank goodness I was able to reach you. This is General Hobbes. We met yesterday at the observatory."

Hank blinked. "Uh—yes, I remember. How can I help you General?"

"I tried calling your colleague—Dr. Marsh, I believe, but that flare knocked all the cell towers offline. Then that reporter, O'Neele, reminded me—you still use a landline."

"Uh—yes, that's right," Hank said.

"Great, well I need you to meet me in your backyard in about fifteen minutes. I'm sending a chopper over to pick you up."

"I don't understand—"

"The flare, man! It has everyone running around like chickens with their heads cut off, and I'm too busy overseeing the military response to continue as the president's science advisor and so he asked me to find a replacement that understood what we're dealing with."

"And you chose—me?" Hank asked, shocked.

"Well, like I said, I tried calling the other guy—no offense, but he's the name I thought of first, but all I had was his cell number. Fortunately, whatever is happening with the flare seems to still allow landlines to work."

"Yes..." Hank said absently, his head spinning. "The electromagnetic—"

"Well," Hobbes said, cutting him off, "I'm sure the president will be very interested in the explanation, but I have to go. The chopper will be there any minute. Oh—I almost forgot. That reporter wanted me to ask you to put on your best suit."

Hank started. "My best suit? What for?"

"For the cameras, obviously. She plans to do an extensive interview with you right after you meet with the president. Dr. Lewis, you're about to become *very* famous."

Part Three

Pricing Psychology and Swamp Hags

The Science

Humans are good at a lot of things that other animals aren't good at: building bridges, creating restaurant chains that combine diverse ethnic cuisines in shocking ways, and creating multimillion dollar sports franchises. But one thing we're *really* good at is convincing other humans to give us money.

People who use illegal means to convince people to give them money are called scam artists. People who use legal means to do it are called advertising departments.

The term *advertise* comes from an ancient Latin term that literally means, "look over there!" This makes sense, as an advertisement's job is to convince you that you should look at its product—ideally in a way that makes you want to purchase it.

Some advertisements do this by presenting an honest and unbiased view of the facts. At least, that's what people tell me. I've never seen such an advertisement. Most advertisements use a clever collection of psychology tricks to convince people that the thing they need most right now just happens to be the very thing the advertiser is selling.

One of the most interesting tricks advertisers use to accomplish this is by trying to convince people that they are lucky.

Imagine one day that you go to the grocery store to check out the price of avocados because you've been dying to try out a new Welsh-Slavic taco fusion recipe. When you walk into the store, you see a bright yellow banner boldly proclaiming:

Tenth Anniversary Sale!
Today only, everything in the store is 10% off!

Now, you might think that a store-wide sale is pretty good at convincing people to spend money, but this type of advertising trick is the work of amateurs.

Meanwhile, in a parallel universe, another version of you walks into the same store with the same avocado needs. Only in this universe, the advertisers are much cleverer. As you enter the store, the manager hurries over to you and hands you what looks like a lottery scratch-off ticket.

"This is your lucky day my friend," he says, flashing a winning smile. "As part of our tenth anniversary sale, a few lucky customers are eligible to win fabulous discounts! Just scratch off this ticket to see if you're a winner!"

Excitedly you scratch off the ticket and read:

Congratulations! You've won 10% off your total purchase! Valid
today only!

In both universes you are presented with the same situation. You can purchase anything in the store for ten percent off. And in both cases, that discount is only valid for today.

So, in which universe are you more likely to spend money?

According to researchers that study marketing and advertising, because the store in the second universe made you feel like you won something, that feeling of "being lucky" will make you much more likely to spend money than a simple store-wide sale, even though both versions of you have access to the same discount.

Researchers from China and Australia recently tried coming up with an Aussie-Mandarin fusion restaurant, but nobody could agree on what should be served for breakfast, so instead they decided to collaborate on developing a better understanding of this "lucky day" phenomenon by creating more complicated shopping scenarios to test their theories.

Imagine that you go out shopping again, this time because you want to buy a new pair of rollerblades. You've heard from a trusted friend that the sport is making a comeback, and because you like to be known as a hipster in the world of recreational sports, you want to act fast before they become too popular again.

Unfortunately, you realize you might have to wait until next week to make your purchase, because you just spent a lot of money on avocados, so money is a little tight. But you decide to go out today and see if there are any good sales on rollerblades— just in case.

In the first universe, you go to a store and see a poster stating:

Buy a pair of rollerblades for $50 and get a free set of steak knives!

Meanwhile, in the parallel universe, parallel you goes to the same store but instead sees this sign:

Buy a set of steak knives for $50 and get a free pair of rollerblades!

In which universe do you think you're more likely to spend the $50?

The researchers showed through a series of experiments that even though both situations give you the exact same deal—obtain rollerblades *and* steak knives for a total of $50, what really matters is whether the promotional part of the bundle includes your *target* purchase.

In other words, you are much more likely to make the purchase if the sign says, "free rollerblades" than if it says, "free steak knives".

What accounts for this behavior? After all, how could a species that invented Mexi-Italian food trucks behave so illogically?

According to surveys given to consumers as part of the experiment, they were motivated by seeing their target item being offered as the free gift, which made them feel lucky. And as other advertising research has shown, when a consumer feels lucky, they are more likely to make a purchase.

Wanting to better understand this phenomenon, researchers conducted a few more experiments and learned that there are two things that can cancel out the advertising power of making someone feel lucky.

The first is if you are forced to buy something that only gives you a *discount* on your target item, you are less likely to be tricked into this deal. So, if the promotion had been:

Buy a set of steak knives for $20 and get 40% off a pair of $50 rollerblades!"

In that situation, you still wouldn't be any more likely to make the purchase than before, even though—assuming my math checks out—you're still getting rollerblades and steak knives for a total of $50.

The second thing that can nullify this fancy "make them feel lucky" technique, is if the consumer knows they are being deliberately marketed to.

So, if you had gone into the store and seen the same promotion as before:

Buy a set of steak knives for $50 and get a free pair of rollerblades!

But then while you're trying on your new blades, you overhear the manager whispering to one of the employees, "Yeah, I know his Mom, and she told us he was interested in rollerblades, so I made that poster a few minutes ago to see if we could convince him to buy some knives from us."

Not only does learning this information cancel out the "feeling lucky" effect, but the research also showed that if you know you're being deliberately marketed to, you'll lose trust in the store and become even less likely to make the purchase than you were before.

While this might seem like cutting-edge research for people working in the advertising and restaurant fusion industries, it's a principle that witches have been exploiting for millennia.

The Fiction

G orna sat on the porch of her hut in the warm afternoon sun, enjoying the satisfying squelch of the bog as it oozed between her gnarled toes. If she curled them just right, she could feel the gritty slime seep beneath her toenails. She closed her eyes and sighed contentedly.

"I don't think this is going to work."

Gorna snorted. Her sister Delna had decided to drop by for a surprise visit, insisting they try and swindle some roadside travelers. *It'll be like old times...* she had said. The trouble was, Gorna didn't care much for the old times, spending hour upon hour arguing with her sister about the best methods to use to ensnare their would-be victims.

The trouble was, people these days were just too cynical. Nobody trusted swamp hags anymore. That was one of the reasons Gorna had decided to leave her home in the west and settle down in the eastern marshes. What she needed was a fresh start, a new perspective on life. And everything had been going beautifully—until her sister Delna had shown up unannounced, knocking on the door of her hut.

Delna wrinkled her long wart-covered nose at the scroll

Gorna had handed to her. "I still think the old ways are best—challenge a traveler to slip their hand in a box and grab what's inside, or make a prince divulge a favorite childhood memory." She sighed wistfully. "Remember the time that old duke came to our hut begging Mama to make some princess fall in love with him, and she offered to do it in exchange for his firstborn son?"

Gorna shook her head in exasperation. "Yes, and then we had to go into hiding for a year while the Duke's army searched for the boy, burning down half the marsh grass in the process."

Delna shrugged. "I don't see the problem."

"The problem, sister, is that those kinds of tactics don't work anymore. Swamp hags have used those tricks for centuries and people have wised up. That's why I wanted to come out here on my own and put my own spin on what it means to be a swamp hag." She stared out across the marsh, her eyes alight with excitement.

Delna frowned. "What's that supposed to mean? You're a hag that lives in a hut in the swamp—conning travelers out of their most prized possessions to teach them a lesson. *That's* what it means to be a swamp hag. At least," she sniffed, "that's how Mama raised us."

"I know all *that*, Delna, that's not what I'm talking about. I just think—look, when's the last time you were able to cajole some prince to slip his hand into a trapped box? Or convince a young maiden to give up her voice?"

"It *has* been a while," Delna admitted grudgingly. "But we don't see much foot traffic back home these days. Ever since the king built his new highway around the dark hills, we don't get many visits."

"Exactly," Gorna said, jumping to her feet and striding over to stand beside Delna. "The king built that highway because of *us*, because of our family. Merchants and travelers don't want to come anywhere near the western swamps—let alone seek us out to grant them boons. But with this..." She took the scroll from Delna's gnarled hand, smiling wickedly. She dropped her voice conspiratorially. "When I first moved out here and bought this hut from an old gnome, he offered to throw in this scroll containing all his best methods for selling his merchandise."

Delna raised an eyebrow. "You bought that scroll from a gnome?"

Gorna shook her head. "No, I received it as a gift when I bought his hut. But that's not important. What *is* important is that I've made a few changes to it since then—gave it the old swamp hag flair. It now has dozens of strategies for convincing people to give up something they love in exchange for their heart's desire. Half the time, they don't even realize they've been tricked."

Delna frowned, glancing at the scroll "Where's the fun in that?"

Gorna waved that away, slipping her lucky scroll into the pocket of her patched robes. "Look, did you hear about Count McFadden's new orphanage?"

"Sure," Delna shrugged. "They say it cost him half his fortune—wait," she eyed her sister who was grinning with sly false modesty. "Are you saying that was *you*?"

Gorna gave a mock curtesy. "The one and only."

"So, your new strategy for trickery is convincing the nobility to help feed the poor?" She raised an eyebrow. "That seems— wrong somehow."

Gorna shook her head. "It's still the same thing, Delna— convincing people to give up their prized possessions to teach them an important lesson. The Count loved his fortune. Now he's learned to love helping the poor too."

"And that's it?" Delna asked skeptically.

"Well..." Gorna hedged, "and in the process he *might* have

managed to convince a local maiden that he was worthy of her hand in marriage."

"Ah-ha!" Delna said, cackling. "I knew self-interest must enter this somewhere, for a moment I'd started to lose all hope in humanity. But this whole charity angle..." She shook her head. "I just don't know."

In the distance, the sound of a crow cawing echoed across the marsh. Gorna whipped her head towards the sound, her eyes alight with excitement. "That's one of my watch birds," she said eagerly. "A traveler approaches. Let's try an experiment."

"What kind of experiment?" Delna asked, frowning.

"It's a variation on the old *three fates routine*, except there's just two of us."

"Obviously," Delna said, rolling her eyes.

Gorna ignored the comment. "We'll both sit here on the deck, and when the traveler approaches, we'll hear his plea, then each make an offer."

"I know how the three fates routine works," Delna said, annoyed. "Honestly, sometimes you forget that *I'm* the older sister."

"Right, but I'll use one of my new methods, and you can use any of the old, traditional ones you like."

Delna opened her mouth to argue but cut off as Gorna held up a hand.

"Here he comes," Gorna said. "Get ready."

Delna sighed, but shuffled over to sit on the deck, letting her legs hang over the side as Gorna had, and dipper her toes into the ooze. "Oh, that's nice."

"Right? I always think—oh, here he comes."

The traveler was on horseback. He was a young man, probably in his early twenties. His fine dress marked him as nobility, or perhaps a wealthy merchant's son. His eyes darted around nervously as he approached, taking in the dilapidated wooden hut and the two swamp hags seated on its deck. His horse whin-nied uneasily, as if it knew what was about to happen and wanted no part of it.

"Greetings traveler," Gorna said pleasantly. "What brings such a handsome young man to our humble cottage?" She used her normal tone of voice. She had been practicing sounding disarmingly coquettish, but she hadn't quite mastered that yet and didn't want to embarrass herself in front of her older sister.

The young man made as if to dismount, then seemed to think better of it and gripped the horse's reins tighter. "I'm here to—" his voice cracked. He flushed slightly, then cleared his throat and tried again. "I'm here today to ask for your assistance in a most urgent matter."

"Indeed?" Gorna asked, flashing a gap-toothed smile. "Please, do go on."

The young man paled slightly, then took a deep breath and plunged onward in a rush. "My father, king of Hebron, has sent me to represent our family in the great tournament of the northern realms."

"Ah yes," Gorna said, "we're familiar with the annual tour-nament. Many great warriors are sure to be present, fighting for glory and honor."

"I suppose," Delna said, sounding bored, "that you wish us to cast a spell upon you, granting you superior strength and speed so you can win the tournament?"

"Er no, actually—"

"Ah, say no more, my young duck," Gorna said, smiling. "I see you come to us weaponless. You wish us to provide you with a powerful blade of great renown, giving you an advantage over the other contestants?"

The young man shook his head. "No, it's nothing like that. You see," he grimaced, seeming embarrassed, then shook his head. "Look, I don't know if you can actually help with this or not, but I thought there would be nothing to lose in coming to ask."

"We'll be the judges of that, my dear," Delna said, leering.

Gorna shushed her. "Go on, young man, we're listening."

He glanced uneasily at Delna. "Well, it's just that, I'm not a warrior, or a prince, or anything like that. I'm not *fighting* in the tournament. Well—not really. You see, my father is a baker of some fame, and normally he would attend, but this year he's taken ill and—"

"And you've come to barter for his life," Delna said, nodding sagely. "Come to ask for a poultice to cure his ailing body?"

The young man shook his head. "No, he'll be fine—just bad indigestion actually, but he doesn't want the bother of the trip this year, so he's sent me in his stead. Only..." he trailed off, looking embarrassed.

"Only what, dear?" Gorna prompted.

"Only, I'm horribly allergic to wheat flour."

Gorna raised an eyebrow. "You're a baker's son and you're—"

"Allergic to wheat flour," the young man finished, nodding morosely. "If I'm anywhere near the stuff my face puffs up and my skin breaks out into these awful—"

"We uh—get the idea," Delna said, making a disgusted face.

"So, you want us to cure your wheat allergy?" Gorna confirmed.

The young man nodded. "If it wouldn't be too much trouble."

Gorna eyed her sister, who just shrugged.

"Very well young man. We'll each present you with an offer. You may then choose to accept either offer or decline both. If you choose to decline both...let me see..." she pursed her lips, eyeing the young man appraisingly. "If you decline both of our offers, you forfeit your horse, which becomes our property to do with as we will. Do you accept these terms?"

The young man swallowed, then nodded. His horse took a skittish step away from the cottage, as if it understood it had just been offered up as collateral. The young man patted its neck reassuringly. "Easy, girl. Everything will be just fine."

"I wouldn't be so presumptuous," Delna cautioned, licking her lips as she eyed the horse.

The horse whinnied indignantly.

"My sister will present the first offer," Gorna said, inclining her head towards Delna.

The young man squared his shoulders, turning to face the other swamp hag. "I'm ready." His voice was almost steady.

Delna smiled mischievously. "My offer is simple, child. Bring me the hair of the woman you love, and in exchange I shall cure you of your—uh, malady."

"Just like that?" The young man said, frowning. "Bring you a lock of hair from—"

Delna held up her hand, interrupting him. "I did not say a *lock* of her hair. I said her hair—I want every strand from tip to scalp. Shave her bald and bring me the hair and you shall be cured."

The young man paled. "But—Esmeralda doesn't even know how I feel about her—I mean, I haven't had time to tell her yet, but what would she say if I asked for her hair?" He shook his head. "It could ruin any chance I'd have with her."

Delna eyed her sister, feeling a moment of desperation. It wasn't so much that she wanted to beat her sister—not really, she just needed her to understand that the old ways were still the best ways.

"As an added incentive," she said, holding up a finger, "I'll give you a tonic guaranteed to make you appear charming to the woman you love for exactly one lunar cycle."

"But what do you want with her hair?" the young man asked, frowning.

"Never you mind," Delna snapped. "Let's just say it'll be a useful way to teach humility to a certain vain young maiden."

"Esmeralda isn't vain," he protested. "She's the most kind and loving—"

"Yes, yes," Delna said, rolling her eyes. "I'm sure she is—probably sings and befriends with woodland creatures too," she added disgustedly. "Do we have a deal or not?"

The young man pursed his lips, considering, then turned to face Gorna. "First, I would hear your sister's offer."

Gorna smiled, one hand fingering the scroll in her pocket. "My offer is one that I believe you'll find both easy to fulfill and advantageous to your future."

"I'm listening," he said warily.

"You see, the reason I was so surprised by your request is because I was cleaning out my hut just this morning and happened to come across a cure for wheat allergies."

"Really?" The young man said, surprised.

"As true as the gold in my teeth," Gorna said, nodding.

Delna snorted, knowing full well her sister's gold tooth was painted.

The young man glanced at her briefly, then turned back to Gorna. "Please, continue."

"Yes, so it seems that this is quite your lucky day. Unfortunately, the code of the swamp hags does require me to make *some* kind of exchange for such potent magic, but I hate to make you pay for something I was just going to end up getting rid of..." her voice trailed off and she tapped her lips as if deep in thought.

"I know," she said brightly. "How about you bring me the young maiden's hair in exchange for a tonic that will make you appear charming to the woman you love for exactly one lunar cycle. Do that, and I'll throw in the wheat allergy cure for free. It really is quite fortunate that you happened by today, because I was just going to throw it out."

The young man looked surprised. "Wow, I think this is the first time I've ever been lucky before. But why are you both so interested in Esmerelda's hair?"

Gorna's face softened and she dropped her voice conspiratorially. "Well, you see it's this little girl from the village. She came by the other day, positively weeping because she'd lost all her hair due to the pox. She begged us to help her grow her hair back, but the truth is, hair is the one thing we have no power over."

"Really?" The young man said, looking surprised.

"As true as I sit before you today," Gorna said. "But the hair of your fair young maiden...well," she shrugged. "It would make a beautiful wig for the poor girl."

The young man found himself nodding. "Yes...and truthfully, Esmerelda's hair would grow back, and if I explained to her that it was meant to help a ease a little girl's suffering..."

To the side, Delna spotted angrily. "Are you kidding me? That's the same deal I made you!"

The young man shook his head. "No, no. I'm sure this is different. I mean—I've never been so lucky before, and if I can do a good deed at the same time..." He set his jaw. "Very well. I accept your offer," he said.

Gorna beamed, then spit into her palm and extended it to the young man. He wrinkled his nose then sighed and spit into his own hand before grasping Gorna's briefly. He then snatched his hand back and wiped it on his trousers.

As the young man rode off on his worse, whistling to

himself, Delna turned to regard her sister, who was watching the boy smugly.

"I suppose you think you're *so* clever," Delna scoffed.

"No, not clever," Gorna said, flourishing her scroll. "Just lucky."

Part Four

Blue Zones and The Demon Dog

The Science

One thing people have in common across all walks of life is a tendency to complain about the life they're walking in. The weather is too hot or too cold, their job is too hard or not fulfilling enough. Their health isn't as good as they'd like, their jeans don't fit the way they used to, and fast-food costs more than it should. They've never won anything, not a beauty contest, a hot dog eating contest, not even a lucky scroll from a gnome. The list of complaints goes on and on.

But, despite how whiney everyone is about their lives, when push comes to shove, most people quite enjoy being alive, and are often very interested in sticking around for as long as possible. Or as several poets, songwriters, and philosophers have quipped over the years, "Everyone wants to go to heaven, but nobody wants to die to get there." Because of this, methods of increasing the length of the mortal journey are always of interest.

So, imagine the excitement generated a few years back, when a group of European scientists published groundbreaking research on longevity—the science of living longer. These scientists had identified certain areas of the world where the average life expectancy was considerably higher than that of surrounding areas. In these areas—which the scientists called Blue Zones—people were regularly living to be close to a hundred years old, and most of those years were spent in relatively good health.

The first question an excited public wanted to know was why they were called *blue* zones. Why not say, green—the color of life? People began to question everything they knew about color theory and design psychology. Was there something special about the color blue that they hadn't realized?

Sales of blue jeans and blue suede shoes hit all-time highs as people set out to create their own personal blue lifestyles. CEOs of major fast-food restaurant chains sat in meetings where top design consultants advocated changing restaurant color schemes from red and yellow to blue and, um—a slightly darker blue.

So, imagine the disappointment when those same research scientists released a statement informing the public that the term "Blue Zone" was chosen simply because they had first used a blue pen to mark the areas of increased longevity on their maps.

After this rather disappointing news, discredited restaurant design consultants had to slip out the back door in the dead of night, and the stock prices for blue jeans plummeted as people returned their blue clothing.

> *Editor's note: I can find no record of these announcements affecting the denim textile industry. However, it should be noted that the internal deliberations of major restaurant chains are a closely guarded secret.*

Even after learning that taking advantage of this longevity research might take more effort than simply purchasing a new wardrobe, the eager public was not deterred. Some people started planning extended vacations to Okinawa, one of the first Blue Zones identified. Others considered painting their houses blue, just in case.

Some people wondered if the scientist's comment about blue pens was all a red herring—or a blue herring. Perhaps they had called these areas Blue Zones because there was a secret, magical substance in the water of these villages that granted its drinkers perpetual youth, and the scientists were now trying to keep it a secret until they could figure out a way to bottle it up and sell it. Suddenly the idea of moving to Okinawa didn't seem so crazy after all.

But as researchers, documentarians, and social media influencers flocked to these Blue Zones and began scrutinizing the lives of those who lived there, the shocking truth became obvious. People in these areas weren't living longer due to the color of their shoes, nor because of the color schemes of local restaurants. And as far as anyone could tell, there wasn't anything different about the water supply.

It turned out that the secret to the long, healthy lives of people living in Blue Zones was that they spent a great deal of time being physically active, avoiding stress, eating locally grown fruits and vegetables, and spending quality time with family and friends.

Once people realized this, they quickly lost interest in the idea. The popular press struggled as they tried to come up with a snappy headline for a breakthrough discovery that boiled down to, "If you want to live longer, stop sitting around all day staring at your phone while eating junk food."

"That seems like an awful lot of work just to live a few more decades," a local man said to a news reporter as he stood in line to return some blue shoes he'd recently purchased. "I mean—if there had been a magic pill or something else you could just buy..." he shrugged, then took a sip of soda from a large blue cup.

Meanwhile, when the mothers of the world heard about all this fuss, they collectively rolled their eyes in exasperation, pointing out that this was exactly what they'd been telling their children for centuries.

But an important thing to always keep in mind when it comes to scientific research is that sometimes there are additional factors you just can't account for—an additional secret or some confounding factor in the data that you miss.

So, while those Blue Zone scientists were busy trekking across jungles and mountain ranges, marking potential hot spots of longevity with their fancy blue pens, perhaps dreaming of the stir their research would cause, they ended up missing one of the most important Blue Zones in the world—the Columbus Park gazebo in Hoboken, New Jersey.

The Fiction

Preston stood just outside the Columbus Park gazebo, thinking hard.

In front of him, a massive genie floated in the air, eyeing him balefully. It was shaped like an extremely ripped man from the waist up, except that its skin shown with a sapphire radiance that bathed the gazebo's interior in an eerie blue light. From the waist down, the creature's form dissolved into smoke, seeming to emanate from the base of the fountain next to the gazebo.

"Any time now," the genie said, glancing up at the moon.

"Don't rush me," Preston said. "It's my last wish; I want to make it a good one."

The genie rolled his eyes. "It *will* be difficult to top those first two you made."

Preston scowled. "Are you making fun of me? I told you I was flustered. It's not every day a genie pops out of a city fountain offering me wishes." He shrugged. "Honestly I thought it might be a joke or something."

"A joke?" The genie sputtered angrily. He swelled to three times his size, towering over Preston. The eerie lights surrounding the genie shifted in hue from blue to red as he thrust his massive face forward until it was inches away from Preston's. "Do I *look* like a joke?"

Preston took a step backwards, swallowing nervously. "N—no, sir."

The genie flashed him a smile of wickedly sharp teeth before shrinking back to his normal size, his skin returning to its original glowing blue color. "Good," he said smugly as he idly examined his fingernails. "Now let's get on with this."

"Why are you so cranky anyway?" Preston asked, eyeing the base of the fountain. "It's not like you're going anywhere."

The genie glared at him. "Oh, I don't know," he said, voice dripping with sarcasm. "Maybe because I'm an ancient and all-powerful being who was just awakened by some punk kid whose first two wishes were for a new set of earbuds and a pack of gum!"

Preston flushed slightly. He really *should* have put more thought into those first two wishes, but when he was nervous, he tended to say the first thing that came to his mind. That was exactly the problem, he thought, cringing as he remembered his last conversation with Laura. If only...his eyes widened.

"Okay, I know what I want to wish for," he said confidently.

The genie sighed. "Let's have it then."

"I wish I were good at sports."

The genie blinked. "Sports? As in—all sports?"

Preston stared at his feet. "Well...track—the running events," he added hurriedly. "You see there's this girl—Laura, and she really likes this kid, Todd, who's on the track team. Last year she went to all his races to cheer for him and kept bragging about how fast he was. The problem is, I'm not much of a runner. But, I figure if you could make me fast enough to make the team and outrun Todd..." He trailed off, shrugging.

The genie rolled his eyes. "Couldn't you just—you know, exercise and eat healthy like everyone else?"

Preston shook his head. "I tried that. It didn't help."

The genie frowned. "Really? How long did you try?"

Preston shrugged. "I don't know, a day or two."

The genie raised an eyebrow. "A day or two?"

"Look, I just don't want to blow it with Laura. Making the team is the only chance I've got, and try outs are only two months away."

The genie's expression softened. "Look kid, I've been around a long time. If you really want to impress this girl—"

Preston glanced at his watch. "Look, I'm sure you're full of old man wisdom and everything, but I didn't wish for a lecture, I just want you to make me run faster."

The genie stared at him, open mouthed. Then he snapped his jaw shut and fixed Preston with a hard gaze.

Preston shivered. "Sorry, I didn't mean—"

The genie held up a hand, smiling wickedly "Okay kid. Your life is about to get your wish." He snapped his fingers and vanished in flash of sapphire light.

The next morning, Preston hurried to get dressed. For the first time since starting high school, he was excited to get to school. When he woke up this morning, he'd thought the whole genie experience had been a dream. Then he'd opened his eyes and saw the pack of gum and new set of earbuds sitting on his nightstand. He'd leapt up, excited. He couldn't wait to get outside and try out his genie-enhanced speed.

He rushed downstairs and reached for his customary breakfast pop tart. Some mornings he took the time to toast the pastry, but today he was in a hurry. He rushed out the door, tearing open the wrapper as he threw his backpack over one shoulder and strode down the sidewalk before turning up the street that led towards the high school. He smiled, absently taking a bite of his pastry then blinked. He was holding a banana.

"What the..." he said frowning. He was sure he'd grabbed a pastry. He didn't even like bananas. Did he have time to go back? He started to turn back towards his house, then froze. A large black dog burst out of the bushes a few yards behind him, surveyed the street briefly, then fixed its eyes on Preston.

"Uh...nice doggie," Preston said. He wasn't sure what kind of dog it was, but it was as tall as his wait and wore a dark blue collar, studded with silver spikes. It's beady black eyes seemed to sparkle as the it stared him down.

"Uh...want a banana?" He tossed the fruit towards the dog. It bounced off its forehead and fell to the sidewalk.

The dog narrowed its eyes at Preston and growled a low, menacing sound that seemed to reverberate in Preston's chest.

He took a nervous step backwards. "Sorry about—yah!" He screamed as the dog gave a loud bark, then broke into a run heading straight for him.

Preston turned and sprinted down the sidewalk, the dog right behind him. He wove past other kids on their way to school, parents, and an old woman pushing a cat in a baby stroller. The dog ignored all of them—even the cat, as it ran after Preston, its snapping jaws just inches away from his feet.

Preston's legs burned as he hurtled up the steps to the school, bursting through the doors, nearly running into a group of confused cheerleaders. He collapsed against a row of lockers, gasping for breath, sweat streaming down his face. He looked back. The dog sat at the base of the steps, watching him for a moment, then got to its feet, shook out its fur, and trotted away around the corner.

Preston sat with eyes closed as he ignored the confused looks and jeers of his schoolmates. What had gotten into that crazy dog? He had just about caught his breath when the first bell rang. Wearily, Preston got to his feet and trudged off to math class.

～

"And then today—it happened again," Preston said wearily. "It chased me all the way here."

"That's insane," his friend, Jackson said, craning his neck to look out the window of their favorite diner. "I don't see it now."

"Of course not," Preston muttered, taking a sip of his water. "Nobody ever sees it but me."

He and Jackson had already ordered and were still waiting for their food. The waiter had already brought them their drinks, though he'd brought Preston a water instead of a soda. But he was too exhausted to complain.

Three times a week for the last two weeks that stupid dog had chased him to school, staying just behind him as he ran the entire way to school. Now it had started chasing him on Saturdays too.

"Have you tried calling the pound?" Jackson asked, swirling the ice around in his soda. The waiter hadn't messed up *his* drink.

Preston nodded. "Twice, but they couldn't find any trace of the stupid thing. I tried the police too, but they just hung up and told me to try a different route to school."

"That's a good idea."

"You'd think so," Preston said with a sigh. "But that dog has it in for me. I tried going out my back door and jumping the fence to the street that runs behind my house, but the dog was

71

right there—waiting for me." He shuddered, remembering. "And that day was worse."

"Why?"

"Because normally it's just over a mile from my house to the school. But going the back way, I had to cut around that construction they're doing on Second—it's all shut down. I had to run all the way over to the Shake Shack before I could cut back towards school; and that dumb dog chasing me the whole time."

Jackson frowned. "Dude, I live on Second, there hasn't been any construction there lately. Are you sure you're feeling okay? Did you say nobody could see this dog except you?"

Preston stared at his friend, who was watching him nervously. "Look, I'm not crazy. I—" he broke off as the waiter approached with their order.

"That's one burger combo with a chocolate shake," the waiter said, setting the plate down in front of Jackson.

"Finally," his friend said. "I'm starving." he wasted no time stuffing a handful of fries into his mouth.

"And one pasta bowl—light on the sauce," the waiter added, setting a dark blue bowl full of pasta in front of Preston.

Preston frowned, staring at the bowl, then turning back to the waiter. "But I didn't order this."

The waiter smiled at him. "Oh, I'm sure that's what you wanted."

Preston frowned. "No, I—" he broke off, staring at the man. "Do I know you from somewhere?"

The waiter raised an eyebrow. "Well, I did just take your order a few minutes ago."

Preston shook his head. "No, I mean—"

"Dude, just eat your pasta," Jackson said. "It's what you ordered."

Preston turned to stare at his friend. "It is?"

Jackson smirked. "I think all that running has done something to your brain. I ordered the burger, you ordered the pasta. I said, *wow, that's a change*, and you said something about being on a new diet—remember?"

Preston stared at his friend, dumbstruck.

Jackson glanced at him nervously. "Maybe you should see a doctor or something."

Preston frowned. Had he really ordered pasta? His stomach growled. Confused and exhausted from spending every other day running from that crazy dog, Preston sighed, picked up his fork and took a bite. It was no bacon cheeseburger, but it wasn't bad.

Preston had started to hate his life. For nearly two months now, that dumb dog—which he had started calling the *demon dog*—had been chasing him. For the last three weeks, it had started chasing him *home* from school too. It was as if it sat waiting for him all day long, biding its time until he could continue tormenting him.

He'd stopped leaving the house on weekends, except to help his mother or grandmother with their errands. It never bothered him when he was out with one of them. In fact, it never bothered him when he was with his friends either. Only when he was walking alone.

His attempts to dissuade the dog had been fruitless. Nobody from animal control took him seriously. The police ignored his calls, and none of his friends had yet to notice the dog, even when it chased him all the way to school. The dog had ignored bribes of bones, raw meat, and even a medium rare steak Preston had prepared in a moment of desperation. He'd tried taking different routes to school, leaving early, leaving late, and inviting friends to walk with him.

No matter what time he left the house, the dog was always there. Whenever he invited friends to accompany him to school, they always had something come up that led to them bailing on him. He'd even tried taking the bus home one day, trying to time his departure from the school so that even with the dog chasing him he'd get to the bus stop just as the bus arrived. But the dog had run faster that day, nearly biting his calf as he ran screaming past the line of people waiting to board the bus.

Preston was starting to think that his friend was right—maybe he *was* losing his mind. But as he glanced uneasily at the pair of earbuds on his dresser, a part of him knew what was happening. That genie had cursed him for calling him an old man. That dog was payback and would likely chase him every day until one of them dropped dead.

He took another bite of his veggie sub and grimaced. His mother had called on her way home from work, offering to pick him up a sandwich. He'd asked for a spicy Italian sub with extra meat. When she'd come home, she'd apologized, telling him the sandwich shop had completely sold out of meat—and dressing. All they'd had left were vegetables and olive oil.

He finished the sub—it actually hadn't been *that* bad—crumbled up the wrapper and tossed it towards the trash can in the corner of his bedroom. It bounced off the rim and fell to the floor. Sighing, he got to his feet, crossed the room and put the crumbled wrapper into the trash. Turning, he caught sight of his wall calendar, it was a month behind.

He removed the pin holding it to the wall and turned the page to March, then paused. Tomorrow's date had been circled, with the word *Tryouts* written in blue ink. What did...then he remembered, track tryouts were tomorrow after school. He'd been so preoccupied trying to think of ways to deal with that insane dog, he'd forgotten all about track.

He wondered if the dog would be there—at track tryouts. Tryouts were outside, but the dog never seemed to bother him when he was at school. He decided anything that let him put off going home would be worth it. Besides, he reasoned—he might run into Laura.

"Okay, this is the final heat for the eight hundred," Coach Greeves called. "I won't lie, we've got a nicely packed varsity roster at this point, but if you're fast enough..." he shrugged.

Preston glanced up at the stands. Laura was there, laughing at something Todd was saying. Preston gritted his teeth and looked at the leaderboard. Todd wasn't the fastest guy on the team, but he was on varsity for both the eight hundred, and sixteen hundred meter. So, Preston had signed up for the same races.

"Runners, take your marks!" the announcer shouted.

Preston moved into position and placed his feet awkwardly on the starting blocks. He wasn't used to running like this.

The starter fired his pistol, and Preston started running. Unfortunately, he stumbled off the blocks while the other boys seemed to explode forward, confident and sure-footed.

Growling in frustration, Preston focused on trying to catch up. He tried to imagine the demon dog on his heels, giving him motivation to run faster. Part of him actually wished the thing were here. Maybe it would chase one of the other kids for a change. But that was too much to hope for.

Preston breathed deeply, surprised to discover that he was halfway through the first lap and wasn't winded at all. Grinning, he dug deeper, pouring every ounce of energy into his run—willing his legs to move faster.

He fought down his own disbelief as he passed two boys, then another—and a fourth. Soon he was just inches behind the leader. Preston pictured the dog in his mind—tried to imagine it right behind him, wearing its blue collar covered in silver spikes, its jagged teeth nipping at his heels as it barked madly. He felt the tempo of his running increase. He drew level with the other runner, then slowly inched ahead of him, then pulled away as the other boy's endurance began to lag.

Suddenly Preston, his body in peak condition after weeks of running from that crazy dog, was in the lead by a wide margin. He lapped two of the slowest runners before flying across the finish line.

Coach Greeves hurried over to Preston, his face alight with amazed excitement. "Son, that was the most amazing run I've seen in years. I think you may have just broken the school record for the eight hundred. If you want a spot on the varsity team, it's yours."

Preston grinned, scanning the crowd of cheering students for a sign of Laura. He saw her, staring at him admiringly, Todd seemingly forgotten—at least for the moment.

Coach Greeves clapped him on the shoulder, still staring at him in wonder "Son, I have to know your secret. What have you been doing to get in such great shape?"

"Thank you, coach," Preston said, "I—" he broke off suddenly. Standing just behind the coach, was a man with blue skin. Nobody else seemed to notice him. He flashed a wicked grin at Preston.

Preston swallowed.

The genie winked at him, then seemed to shrink in on himself, transforming into a large black dog with a blue collar. The dog barked once, then turned and trotted off.

Preston's eyes followed it for a few seconds. He felt slightly dizzy.

"You all right, kid?" Coach Greeves asked, frowning. "Need some water?"

Preston shook himself. "Sorry, coach. I was just thinking about your question."

"And?" Coach Greeves prompted. "What's the magic formula to your success?"

Preston shrugged. "No magic—just diet and exercise."

References

Cryptic Evolution and Ancient Warriors

Gibson, G., & Dworkin, I. (2004). Uncovering cryptic genetic variation. Nature Reviews Genetics, 5(9), 681–690. https://doi.org/10.1038/nrg1426

Solar Scientists and Vintage Glory

Wang, J. (2024). Solar and heliospheric observatory (Soho): Living with a restless star. In J. Wang (Ed.), Eye Beyond the Sky: 27 Telescopes and Space Probes, from Hooker to JWST (pp. 275–288). Springer Nature. https://doi.org/10.1007/978-981-99-9818-0_19

Lucky Days and Swamp Hags

Liu, M. W., Wei, C., Yang, L., & Keh, H. T. (2022). Feeling lucky: How framing the target product as a free gift enhances purchase intention. International Journal of Research in Marketing, 39(2), 349–363. https://doi.org/10.1016/j.ijresmar.2021.07.001

Blue Zones and Magical Fitness

Poulain, M., Herm, A., & Pes, G. (2013). The Blue Zones: Areas of exceptional longevity around the world. Vienna Yearbook of Population Research, 11, 87–108. https://www.jstor.org/stable/43050798

About the Author

LEE FALIN is a best-selling middle-grade and young adult author, and the host of the My Cousin Jane podcast. Lee has a PhD in Genetics, but these days he writes more fiction than research.

Lee is married to a wonderful woman he met while they were both serving as volunteer missionaries in Brazil. They have five amazing children, a shocking number of which are now adults.